Math Concept Reader

Counting in the City

by Jean Sharp
Illustrations by Lorin Walter

Copyright © Gareth Stevens, Inc. All rights reserved.

Developed for Harcourt, Inc., by Gareth Stevens, Inc.
This edition published by Harcourt, Inc., by agreement with Gareth Stevens, Inc. No part of this publication may be reproduced or transmitted in any form or by any means, electronic or mechanical, including photocopy, recording, or any information storage and retrieval system, without permission in writing from the copyright holder.

Requests for permission to make copies of any part of the work should be addressed to Permissions Department, Gareth Stevens, Inc., 330 West Olive Street, Suite 100, Milwaukee, Wisconsin 53212. Fax: 414-332-3567.

HARCOURT and the Harcourt Logo are trademarks of Harcourt, Inc., registered in the United States of America and/or other jurisdictions.

Printed in the United States of America

ISBN 13: 978-0-15-360163-7
ISBN 10: 0-15-360163-9

1 2 3 4 5 6 7 8 9 10 039 16 15 14 13 12 11 10 09 08 07

Mom and I take a train to the city.
This is my first trip on a train.

The city is a very busy place.
We will find many things to count.

We walk down the city sidewalk.
I count 9 tall buildings in a row.

I count 7 streetlights.
They stand tall and straight.

I count 5 yellow taxis driving by.
People ride in them around the city.

I count 2 red buses on the street.
The buses travel from stop to stop.

We see many people in the city.
We see 3 firefighters.
They work at the firehouse.

There are 4 police officers.
They are at the police station.
The station is near the firehouse.

I see 10 children playing.
They run and jump together.
They have fun playing in the city.

I see 6 teachers with the children.
They watch the children play.
They make sure no one gets hurt.

We go to the park in the city.
We count 17 balloons on strings.
The balloons are different colors.

We count 13 dogs and 0 cats.
The dogs like walking in the park.
A long brown dog wags its tail.

We sit by the pond to eat lunch.
We see 16 boats on the water.
People row the boats with oars.

We count 20 ducks.

We see a mother duck swimming by.

She swims with her ducklings.

We see many stores in the city.
I count 12 muffins at a bakery.

A bookstore is next to the bakery.
I count 11 books in the window.

We see a fruit stand on the corner.
I count 8 green apples in a basket.

We see a flower shop next.
I count 14 red flowers at the shop.

It is night now in the city.
I count lights in 30 windows.

I look up above me.
I count 24 stars in the sky.

There is 1 train on the track.
Mom and I ride it home.

It is fun to count in the city.
I hope we come back again soon.

Glossary

bakery

city

firefighter

fruit stand

muffin

police officer

taxi